**TRAIN MY BRAIN**
Global

# ADULT **MATH** ACTIVITY B

## 3 Digit Problems – Numbers 100 t

D1454401

# Contents

Welcome ............................................................ 03

Addition ............................................................ 04

Subtraction ....................................................... 24

Multiplication ................................................... 44

Division ............................................................ 64

Bonus: Next Book Sample – 4-Digit Problems ............ 84

Bonus: Mixed Operations Sample ........................... 85

Bonus: Multi-Step Long Math Sample ...................... 86

Bonus: Algebra Book Sample ................................ 87

Answers ............................................................ 88

Other Math Books & FREE Book Club with Discounts ........ 108

## Contact the Author

My name is Grace Hartford, the author of this book.

You can email me at **grace@smile.ws**

I have spent countless hours on this book
to create the best experience for my readers.

I would love to hear any feedback. Alternatively, if you have any issues,
please email me, and I will sort them out.

# Other Math Books

## Arithmetic Series

Practice arithmetic problems
Visit: **smile.ws/pmа9**

## Multi-Step Math Series

Multi-step math problems
Visit: **smile.ws/pmm9**

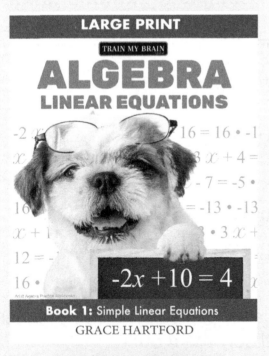

## Algebra Series

5 types of problems
Visit: **smile.ws/pmx9**

**Turn to pages 84-87 for more puzzle books and bonus samples!**

Join our free club for discount coupons on future books!

Visit: **smile.ws/cmа9**

## Welcome

I hope you enjoy this book of math problems. They are a fun way to train your brain by keeping your mind active!

## Extra Wide Margins – Rip out the Pages!

Every page of math problems in this book has extra-wide margins. So you can easily rip out the pages, which can make it more convenient to solve the problems.

## More Math Books Available: 7 in Total!

**Book 1:** Two-digit math problems      (Numbers 10-99)

**Book 2:** Three-digit math problems   (Numbers 100-999)

**Book 3:** Four-digit math problems     (Numbers 1000-9999)

**Book 4:** Mixed digit math problems   (Numbers 10-999)

**Book 5:** Mixed digit math problems   (Numbers 10-9999)

**Book 6:** Mixed problems

**Book 7:** Mixed operations (Numbers 1-99)

Collect them all for maximum challenge! (This book is **Book 2**)

Enjoy the book. Love,

*Grace xxx*

Grace Hartford – Founder of TRAIN MY BRAIN

**Turn to pages 84-87 for bonus samples of my other books!**

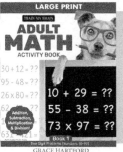

# Addition

1)  360
  + 264
  _____

2)  803
  + 416
  _____

3)  501
  + 465
  _____

4)  646
  + 644
  _____

5)  613
  + 933
  _____

6)  634
  + 205
  _____

7)  204
  + 946
  _____

8)  713
  + 329
  _____

9)  434
  + 114
  _____

10)  648
  + 498
  _____

11)  657
  + 892
  _____

12)  604
  + 693
  _____

13)  721
  + 226
  _____

14)  904
  + 474
  _____

15)  752
  + 623
  _____

# Addition

Rip out the pages!

16)  352
   + 868
   _____

17)  573
   + 213
   _____

18)  478
   + 753
   _____

19)  133
   + 725
   _____

20)  199
   + 206
   _____

21)  843
   + 933
   _____

22)  482
   + 688
   _____

23)  862
   + 251
   _____

24)  305
   + 558
   _____

25)  389
   + 592
   _____

26)  844
   + 169
   _____

27)  777
   + 171
   _____

28)  412
   + 504
   _____

29)  133
   + 654
   _____

30)  263
   + 155
   _____

# Addition

31)    874
    + 599
_____

32)    183
    + 458
_____

33)    381
    + 798
_____

34)    993
    + 288
_____

35)    449
    + 699
_____

36)    161
    + 366
_____

37)    714
    + 966
_____

38)    941
    + 272
_____

39)    881
    + 279
_____

40)    515
    + 335
_____

41)    454
    + 847
_____

42)    238
    + 340
_____

43)    267
    + 526
_____

44)    332
    + 642
_____

45)    875
    + 471
_____

# Addition

46)  529
   + 200
   _____

47)  890
   + 291
   _____

48)  482
   + 715
   _____

49)  458
   + 264
   _____

50)  806
   + 590
   _____

51)  636
   + 209
   _____

52)  354
   + 511
   _____

53)  700
   + 282
   _____

54)  586
   + 777
   _____

55)  465
   + 346
   _____

56)  655
   + 825
   _____

57)  887
   + 210
   _____

58)  679
   + 414
   _____

59)  882
   + 168
   _____

60)  805
   + 759
   _____

# Addition

61)  190
   + 648
   _____

62)  370
   + 665
   _____

63)  826
   + 648
   _____

64)  684
   + 603
   _____

65)  986
   + 866
   _____

66)  192
   + 780
   _____

67)  696
   + 964
   _____

68)  386
   + 362
   _____

69)  333
   + 595
   _____

70)  837
   + 112
   _____

71)  340
   + 646
   _____

72)  322
   + 383
   _____

73)  518
   + 684
   _____

74)  612
   + 148
   _____

75)  595
   + 198
   _____

76)  783
    + 919
    —————

77)  857
    + 722
    —————

78)  214
    + 392
    —————

79)  380
    + 266
    —————

80)  162
    + 773
    —————

81)  532
    + 953
    —————

82)  517
    + 269
    —————

83)  286
    + 449
    —————

84)  618
    + 729
    —————

85)  966
    + 434
    —————

86)  204
    + 864
    —————

87)  148
    + 160
    —————

88)  635
    + 662
    —————

89)  225
    + 376
    —————

90)  460
    + 274
    —————

# Addition

91)  294
   + 764
   _____

92)  768
   + 510
   _____

93)  107
   + 843
   _____

94)  459
   + 447
   _____

95)  917
   + 246
   _____

96)  667
   + 206
   _____

97)  986
   + 925
   _____

98)  441
   + 889
   _____

99)  727
   + 761
   _____

100) 460
   + 596
   _____

101) 820
   + 271
   _____

102) 649
   + 922
   _____

103) 400
   + 286
   _____

104) 803
   + 158
   _____

105) 477
   + 719
   _____

# Addition

106) 956
    + 330
    ―――――

107) 438
    + 408
    ―――――

108) 399
    + 709
    ―――――

109) 652
    + 816
    ―――――

110) 283
    + 378
    ―――――

111) 784
    + 742
    ―――――

112) 379
    + 748
    ―――――

113) 190
    + 168
    ―――――

114) 978
    + 481
    ―――――

115) 316
    + 881
    ―――――

116) 529
    + 605
    ―――――

117) 335
    + 616
    ―――――

118) 312
    + 985
    ―――――

119) 735
    + 854
    ―――――

120) 450
    + 829
    ―――――

# Addition

121) 607
   + 725
   _____

122) 689
   + 127
   _____

123) 500
   + 420
   _____

124) 522
   + 869
   _____

125) 125
   + 810
   _____

126) 981
   + 629
   _____

127) 296
   + 651
   _____

128) 707
   + 393
   _____

129) 686
   + 131
   _____

130) 213
   + 247
   _____

131) 779
   + 945
   _____

132) 262
   + 482
   _____

133) 994
   + 739
   _____

134) 552
   + 813
   _____

135) 928
   + 240
   _____

# Addition

**136)** 919
+ 866
_____

**137)** 914
+ 583
_____

**138)** 362
+ 795
_____

**139)** 723
+ 541
_____

**140)** 955
+ 871
_____

**141)** 703
+ 501
_____

**142)** 138
+ 122
_____

**143)** 871
+ 871
_____

**144)** 659
+ 103
_____

**145)** 525
+ 513
_____

**146)** 539
+ 462
_____

**147)** 247
+ 345
_____

**148)** 207
+ 320
_____

**149)** 222
+ 815
_____

**150)** 114
+ 277
_____

# Addition

151) 147
    + 180
    _____

152) 772
    + 743
    _____

153) 573
    + 840
    _____

154) 287
    + 587
    _____

155) 317
    + 673
    _____

156) 268
    + 960
    _____

157) 771
    + 249
    _____

158) 472
    + 481
    _____

159) 923
    + 713
    _____

160) 239
    + 865
    _____

161) 143
    + 208
    _____

162) 731
    + 757
    _____

163) 518
    + 608
    _____

164) 572
    + 617
    _____

165) 818
    + 805
    _____

# Addition

166) 745
+ 994
_____

167) 324
+ 818
_____

168) 799
+ 873
_____

169) 558
+ 689
_____

170) 884
+ 115
_____

171) 441
+ 919
_____

172) 210
+ 346
_____

173) 280
+ 474
_____

174) 834
+ 123
_____

175) 530
+ 415
_____

176) 521
+ 418
_____

177) 886
+ 412
_____

178) 659
+ 257
_____

179) 822
+ 654
_____

180) 732
+ 845
_____

# Addition

181) 317
+ 585
_____

182) 553
+ 191
_____

183) 589
+ 702
_____

184) 302
+ 199
_____

185) 637
+ 771
_____

186) 183
+ 932
_____

187) 584
+ 816
_____

188) 671
+ 434
_____

189) 193
+ 135
_____

190) 889
+ 500
_____

191) 697
+ 971
_____

192) 296
+ 950
_____

193) 479
+ 597
_____

194) 338
+ 482
_____

195) 538
+ 477
_____

# Addition

**196)** 586
+ 674
_____

**197)** 540
+ 581
_____

**198)** 100
+ 733
_____

**199)** 110
+ 845
_____

**200)** 485
+ 839
_____

**201)** 304
+ 557
_____

**202)** 790
+ 361
_____

**203)** 164
+ 809
_____

**204)** 228
+ 452
_____

**205)** 139
+ 378
_____

**206)** 404
+ 449
_____

**207)** 196
+ 203
_____

**208)** 780
+ 994
_____

**209)** 579
+ 900
_____

**210)** 885
+ 693
_____

# Addition

211) 775
    + 720
    _____

212) 578
    + 796
    _____

213) 893
    + 833
    _____

214) 632
    + 857
    _____

215) 138
    + 571
    _____

216) 320
    + 616
    _____

217) 711
    + 675
    _____

218) 952
    + 950
    _____

219) 311
    + 630
    _____

220) 346
    + 622
    _____

221) 882
    + 739
    _____

222) 843
    + 350
    _____

223) 570
    + 441
    _____

224) 876
    + 921
    _____

225) 755
    + 263
    _____

# Addition

226) 347
+ 477
_____

227) 712
+ 128
_____

228) 363
+ 917
_____

229) 186
+ 119
_____

230) 595
+ 828
_____

231) 624
+ 231
_____

232) 257
+ 877
_____

233) 999
+ 275
_____

234) 626
+ 192
_____

235) 570
+ 429
_____

236) 148
+ 406
_____

237) 550
+ 534
_____

238) 500
+ 637
_____

239) 738
+ 948
_____

240) 940
+ 142
_____

# Addition

241) 176
   + 838
   _____

242) 624
   + 801
   _____

243) 742
   + 351
   _____

244) 526
   + 538
   _____

245) 168
   + 900
   _____

246) 322
   + 586
   _____

247) 696
   + 579
   _____

248) 761
   + 274
   _____

249) 640
   + 326
   _____

250) 342
   + 970
   _____

251) 415
   + 961
   _____

252) 665
   + 684
   _____

253) 860
   + 384
   _____

254) 115
   + 388
   _____

255) 232
   + 973
   _____

# Addition

256) 229
+ 742
_____

257) 935
+ 601
_____

258) 752
+ 161
_____

259) 257
+ 459
_____

260) 635
+ 569
_____

261) 604
+ 869
_____

262) 846
+ 153
_____

263) 933
+ 912
_____

264) 183
+ 690
_____

265) 264
+ 357
_____

266) 264
+ 651
_____

267) 738
+ 424
_____

268) 718
+ 859
_____

269) 934
+ 516
_____

270) 307
+ 401
_____

# Addition

271) 842
+ 220
_____

272) 634
+ 104
_____

273) 538
+ 935
_____

274) 105
+ 374
_____

275) 519
+ 271
_____

276) 440
+ 734
_____

277) 169
+ 140
_____

278) 132
+ 350
_____

279) 913
+ 909
_____

280) 674
+ 612
_____

281) 896
+ 337
_____

282) 209
+ 113
_____

283) 501
+ 646
_____

284) 656
+ 442
_____

285) 812
+ 552
_____

# Addition

286) 181
+ 958
_____

287) 468
+ 788
_____

288) 996
+ 420
_____

289) 912
+ 743
_____

290) 766
+ 735
_____

291) 601
+ 778
_____

292) 797
+ 718
_____

293) 602
+ 445
_____

294) 112
+ 668
_____

295) 398
+ 611
_____

296) 549
+ 112
_____

297) 555
+ 512
_____

298) 389
+ 396
_____

299) 281
+ 562
_____

300) 254
+ 155
_____

# Subtraction

1)   909
   - 553
   _____

2)   849
   - 319
   _____

3)   591
   - 301
   _____

4)   737
   - 725
   _____

5)   629
   - 334
   _____

6)   836
   - 647
   _____

7)   540
   - 491
   _____

8)   904
   - 462
   _____

9)   310
   - 199
   _____

10)  954
   - 371
   _____

11)  948
   - 214
   _____

12)  932
   - 205
   _____

13)  753
   - 212
   _____

14)  676
   - 586
   _____

15)  825
   - 662
   _____

# Subtraction

16) 275
    - 234
    _____

17) 942
    - 827
    _____

18) 965
    - 567
    _____

19) 764
    - 351
    _____

20) 513
    - 211
    _____

21) 677
    - 631
    _____

22) 911
    - 136
    _____

23) 574
    - 206
    _____

24) 943
    - 413
    _____

25) 795
    - 469
    _____

26) 234
    - 157
    _____

27) 625
    - 422
    _____

28) 438
    - 403
    _____

29) 632
    - 109
    _____

30) 964
    - 429
    _____

# Subtraction

31)   805
  - 714
_____

32)   651
  - 425
_____

33)   353
  - 334
_____

34)   595
  - 300
_____

35)   455
  - 345
_____

36)   539
  - 137
_____

37)   556
  - 121
_____

38)   617
  - 140
_____

39)   437
  - 257
_____

40)   684
  - 499
_____

41)   645
  - 581
_____

42)   820
  - 728
_____

43)   786
  - 314
_____

44)   790
  - 634
_____

45)   489
  - 210
_____

# Subtraction

46)  771
   - 665
   _____

47)  299
   - 158
   _____

48)  706
   - 585
   _____

49)  990
   - 943
   _____

50)  237
   - 126
   _____

51)  535
   - 322
   _____

52)  969
   - 351
   _____

53)  700
   - 533
   _____

54)  413
   - 365
   _____

55)  509
   - 414
   _____

56)  859
   - 624
   _____

57)  832
   - 700
   _____

58)  579
   - 538
   _____

59)  464
   - 207
   _____

60)  685
   - 385
   _____

# Subtraction

61)  773
  - 719
  _____

62)  845
  - 170
  _____

63)  793
  - 493
  _____

64)  804
  - 182
  _____

65)  471
  - 183
  _____

66)  478
  - 100
  _____

67)  797
  - 632
  _____

68)  555
  - 331
  _____

69)  725
  - 303
  _____

70)  985
  - 323
  _____

71)  908
  - 655
  _____

72)  403
  - 385
  _____

73)  423
  - 350
  _____

74)  587
  - 191
  _____

75)  774
  - 170
  _____

# Subtraction

76)  694
   - 652
   ———

77)  798
   - 321
   ———

78)  630
   - 151
   ———

79)  963
   - 390
   ———

80)  968
   - 553
   ———

81)  581
   - 266
   ———

82)  773
   - 371
   ———

83)  778
   - 499
   ———

84)  802
   - 689
   ———

85)  959
   - 121
   ———

86)  351
   - 342
   ———

87)  963
   - 303
   ———

88)  858
   - 603
   ———

89)  211
   - 210
   ———

90)  981
   - 605
   ———

# Subtraction

91)  464
  - 433
  _____

92)  949
  - 263
  _____

93)  926
  - 448
  _____

94)  928
  - 433
  _____

95)  432
  - 152
  _____

96)  650
  - 415
  _____

97)  861
  - 534
  _____

98)  912
  - 572
  _____

99)  850
  - 493
  _____

100) 727
  - 623
  _____

101) 782
  - 385
  _____

102) 714
  - 172
  _____

103) 863
  - 292
  _____

104) 704
  - 579
  _____

105) 148
  - 145
  _____

# Subtraction

106) 759
- 358
———

107) 352
- 285
———

108) 422
- 247
———

109) 870
- 615
———

110) 585
- 161
———

111) 552
- 206
———

112) 445
- 437
———

113) 792
- 392
———

114) 605
- 463
———

115) 930
- 723
———

116) 884
- 172
———

117) 327
- 261
———

118) 374
- 243
———

119) 619
- 363
———

120) 932
- 353
———

# Subtraction

121) 684
   - 235
   _____

122) 735
   - 422
   _____

123) 736
   - 598
   _____

124) 621
   - 328
   _____

125) 967
   - 389
   _____

126) 969
   - 505
   _____

127) 504
   - 105
   _____

128) 761
   - 234
   _____

129) 908
   - 831
   _____

130) 976
   - 170
   _____

131) 810
   - 299
   _____

132) 400
   - 185
   _____

133) 724
   - 521
   _____

134) 455
   - 155
   _____

135) 323
   - 242
   _____

# Subtraction

136) 788
- 608
_____

137) 614
- 378
_____

138) 833
- 665
_____

139) 730
- 703
_____

140) 244
- 157
_____

141) 606
- 585
_____

142) 833
- 450
_____

143) 843
- 842
_____

144) 809
- 403
_____

145) 621
- 196
_____

146) 815
- 720
_____

147) 648
- 383
_____

148) 274
- 271
_____

149) 860
- 360
_____

150) 227
- 201
_____

# Subtraction

151) 489
    - 259
_____

152) 516
    - 106
_____

153) 835
    - 769
_____

154) 931
    - 751
_____

155) 991
    - 858
_____

156) 770
    - 133
_____

157) 582
    - 427
_____

158) 777
    - 179
_____

159) 394
    - 256
_____

160) 939
    - 496
_____

161) 761
    - 429
_____

162) 713
    - 315
_____

163) 551
    - 234
_____

164) 327
    - 207
_____

165) 291
    - 185
_____

# Subtraction

166) 619
- 574
_____

167) 390
- 310
_____

168) 653
- 374
_____

169) 956
- 904
_____

170) 683
- 572
_____

171) 822
- 334
_____

172) 931
- 856
_____

173) 933
- 839
_____

174) 824
- 149
_____

175) 721
- 157
_____

176) 936
- 900
_____

177) 810
- 524
_____

178) 465
- 217
_____

179) 755
- 501
_____

180) 278
- 257
_____

# Subtraction

181) 986
  - 193
  _____

182) 743
  - 558
  _____

183) 626
  - 121
  _____

184) 997
  - 328
  _____

185) 971
  - 275
  _____

186) 451
  - 446
  _____

187) 322
  - 106
  _____

188) 813
  - 631
  _____

189) 688
  - 549
  _____

190) 909
  - 877
  _____

191) 649
  - 267
  _____

192) 998
  - 209
  _____

193) 690
  - 680
  _____

194) 520
  - 215
  _____

195) 222
  - 196
  _____

# Subtraction

196) 882
  - 570
  _____

197) 724
  - 224
  _____

198) 338
  - 297
  _____

199) 943
  - 722
  _____

200) 959
  - 956
  _____

201) 675
  - 146
  _____

202) 735
  - 196
  _____

203) 820
  - 452
  _____

204) 528
  - 115
  _____

205) 973
  - 716
  _____

206) 658
  - 147
  _____

207) 908
  - 275
  _____

208) 799
  - 203
  _____

209) 626
  - 101
  _____

210) 733
  - 161
  _____

# Subtraction

211) 852
    - 491
    _____

212) 180
    - 117
    _____

213) 855
    - 378
    _____

214) 794
    - 652
    _____

215) 803
    - 127
    _____

216) 558
    - 358
    _____

217) 797
    - 709
    _____

218) 737
    - 298
    _____

219) 959
    - 341
    _____

220) 592
    - 333
    _____

221) 900
    - 198
    _____

222) 453
    - 335
    _____

223) 967
    - 775
    _____

224) 958
    - 946
    _____

225) 708
    - 104
    _____

# Subtraction

226) 374
  - 347
  _____

227) 851
  - 210
  _____

228) 605
  - 511
  _____

229) 438
  - 369
  _____

230) 604
  - 364
  _____

231) 713
  - 659
  _____

232) 850
  - 124
  _____

233) 527
  - 145
  _____

234) 228
  - 109
  _____

235) 655
  - 380
  _____

236) 644
  - 589
  _____

237) 478
  - 271
  _____

238) 688
  - 324
  _____

239) 943
  - 484
  _____

240) 583
  - 201
  _____

# Subtraction

241) 892
   - 558
   _____

242) 750
   - 690
   _____

243) 564
   - 434
   _____

244) 622
   - 101
   _____

245) 766
   - 355
   _____

246) 934
   - 173
   _____

247) 553
   - 387
   _____

248) 862
   - 387
   _____

249) 930
   - 780
   _____

250) 777
   - 600
   _____

251) 956
   - 902
   _____

252) 514
   - 257
   _____

253) 849
   - 457
   _____

254) 977
   - 404
   _____

255) 774
   - 544
   _____

# Subtraction

256) 642
   - 206
   _____

257) 904
   - 517
   _____

258) 695
   - 614
   _____

259) 935
   - 104
   _____

260) 563
   - 532
   _____

261) 846
   - 365
   _____

262) 796
   - 111
   _____

263) 905
   - 288
   _____

264) 231
   - 151
   _____

265) 852
   - 503
   _____

266) 271
   - 202
   _____

267) 987
   - 907
   _____

268) 270
   - 120
   _____

269) 835
   - 155
   _____

270) 767
   - 282
   _____

# Subtraction

271) 657
  - 126
  _____

272) 919
  - 520
  _____

273) 536
  - 536
  _____

274) 659
  - 334
  _____

275) 653
  - 190
  _____

276) 469
  - 232
  _____

277) 907
  - 675
  _____

278) 848
  - 775
  _____

279) 404
  - 349
  _____

280) 555
  - 255
  _____

281) 537
  - 301
  _____

282) 405
  - 402
  _____

283) 913
  - 660
  _____

284) 476
  - 225
  _____

285) 959
  - 377
  _____

# Subtraction

286) 511
 - 141
_____

287) 832
 - 446
_____

288) 516
 - 485
_____

289) 912
 - 667
_____

290) 643
 - 421
_____

291) 942
 - 261
_____

292) 436
 - 354
_____

293) 601
 - 175
_____

294) 302
 - 171
_____

295) 803
 - 508
_____

296) 276
 - 250
_____

297) 273
 - 195
_____

298) 605
 - 398
_____

299) 430
 - 362
_____

300) 878
 - 344
_____

# Multiplication

1)     288
   x 245
_____

2)     149
   x 883
_____

3)     863
   x 853
_____

4)     382
   x 593
_____

5)     114
   x 602
_____

6)     721
   x 717
_____

7)     228
   x 373
_____

8)     592
   x 900
_____

9)     165
   x 861
_____

10)    546
   x 239
_____

11)    748
   x 324
_____

12)    624
   x 835
_____

13)    673
   x 148
_____

14)    873
   x 743
_____

15)    625
   x 518
_____

# Multiplication

16) 539
x 439
_____

17) 595
x 131
_____

18) 560
x 610
_____

19) 725
x 348
_____

20) 265
x 590
_____

21) 470
x 646
_____

22) 486
x 136
_____

23) 244
x 142
_____

24) 254
x 696
_____

25) 387
x 495
_____

26) 647
x 878
_____

27) 921
x 901
_____

28) 793
x 257
_____

29) 656
x 484
_____

30) 277
x 241
_____

# Multiplication

31)   579
   x 102
___

32)   160
   x 922
___

33)   153
   x 866
___

34)   468
   x 799
___

35)   502
   x 859
___

36)   464
   x 855
___

37)   511
   x 975
___

38)   580
   x 534
___

39)   698
   x 240
___

40)   545
   x 653
___

41)   359
   x 185
___

42)   873
   x 997
___

43)   242
   x 123
___

44)   740
   x 559
___

45)   724
   x 181
___

# Multiplication

46) 494
x 557
_____

47) 494
x 614
_____

48) 216
x 225
_____

49) 908
x 848
_____

50) 444
x 226
_____

51) 947
x 237
_____

52) 462
x 450
_____

53) 657
x 738
_____

54) 950
x 230
_____

55) 286
x 751
_____

56) 310
x 435
_____

57) 486
x 391
_____

58) 230
x 367
_____

59) 967
x 246
_____

60) 251
x 113
_____

# Multiplication

61)  177
  x 894
  _____

62)  783
  x 879
  _____

63)  642
  x 197
  _____

64)  406
  x 162
  _____

65)  863
  x 965
  _____

66)  710
  x 623
  _____

67)  811
  x 184
  _____

68)  488
  x 993
  _____

69)  525
  x 158
  _____

70)  101
  x 605
  _____

71)  593
  x 241
  _____

72)  443
  x 876
  _____

73)  462
  x 730
  _____

74)  241
  x 996
  _____

75)  841
  x 432
  _____

# Multiplication

76) 342
    x 885

77) 669
    x 197

78) 371
    x 366

79) 240
    x 173

80) 964
    x 710

81) 527
    x 276

82) 324
    x 813

83) 845
    x 757

84) 304
    x 872

85) 662
    x 518

86) 922
    x 220

87) 935
    x 343

88) 248
    x 956

89) 453
    x 105

90) 518
    x 886

# Multiplication

91) 795
   x 603
   _____

92) 400
   x 919
   _____

93) 392
   x 466
   _____

94) 538
   x 342
   _____

95) 774
   x 841
   _____

96) 502
   x 323
   _____

97) 449
   x 820
   _____

98) 736
   x 103
   _____

99) 822
   x 837
   _____

100) 824
    x 534
    _____

101) 135
    x 362
    _____

102) 184
    x 437
    _____

103) 425
    x 909
    _____

104) 799
    x 917
    _____

105) 670
    x 807
    _____

# Multiplication

106) 481
x 523
_____

107) 610
x 337
_____

108) 349
x 342
_____

109) 725
x 597
_____

110) 631
x 632
_____

111) 632
x 868
_____

112) 399
x 947
_____

113) 311
x 413
_____

114) 484
x 270
_____

115) 396
x 677
_____

116) 195
x 755
_____

117) 342
x 200
_____

118) 686
x 174
_____

119) 645
x 905
_____

120) 733
x 410
_____

# Multiplication

121) 341
x 707

122) 864
x 323

123) 488
x 771

124) 949
x 497

125) 966
x 760

126) 391
x 219

127) 873
x 291

128) 682
x 133

129) 671
x 657

130) 813
x 802

131) 527
x 885

132) 589
x 281

133) 649
x 523

134) 469
x 146

135) 553
x 153

# Multiplication

136) 535
x 185
_____

137) 692
x 511
_____

138) 550
x 645
_____

139) 406
x 973
_____

140) 378
x 965
_____

141) 186
x 847
_____

142) 287
x 519
_____

143) 136
x 448
_____

144) 226
x 256
_____

145) 142
x 717
_____

146) 931
x 536
_____

147) 154
x 699
_____

148) 866
x 627
_____

149) 130
x 476
_____

150) 581
x 107
_____

# Multiplication

151) 746
x 584
_____

152) 872
x 994
_____

153) 977
x 135
_____

154) 651
x 896
_____

155) 107
x 367
_____

156) 981
x 823
_____

157) 118
x 640
_____

158) 141
x 753
_____

159) 271
x 400
_____

160) 343
x 863
_____

161) 689
x 748
_____

162) 380
x 166
_____

163) 168
x 661
_____

164) 266
x 237
_____

165) 633
x 504
_____

# Multiplication

166) 229
   x 361
_____

167) 887
   x 798
_____

168) 673
   x 124
_____

169) 713
   x 734
_____

170) 644
   x 951
_____

171) 419
   x 726
_____

172) 910
   x 999
_____

173) 393
   x 286
_____

174) 345
   x 810
_____

175) 441
   x 173
_____

176) 852
   x 450
_____

177) 189
   x 867
_____

178) 955
   x 344
_____

179) 613
   x 269
_____

180) 507
   x 908
_____

# Multiplication

181) 569
x 251
_____

182) 394
x 922
_____

183) 522
x 817
_____

184) 717
x 900
_____

185) 673
x 828
_____

186) 647
x 302
_____

187) 992
x 589
_____

188) 580
x 160
_____

189) 302
x 324
_____

190) 838
x 524
_____

191) 554
x 310
_____

192) 634
x 886
_____

193) 733
x 198
_____

194) 210
x 178
_____

195) 445
x 155
_____

# Multiplication

196) 727
x 177

197) 862
x 957

198) 533
x 877

199) 591
x 973

200) 425
x 907

201) 315
x 297

202) 377
x 912

203) 825
x 258

204) 978
x 772

205) 832
x 445

206) 895
x 359

207) 810
x 712

208) 459
x 134

209) 728
x 379

210) 373
x 527

# Multiplication

211) 151
x 943
_____

212) 676
x 956
_____

213) 332
x 380
_____

214) 214
x 574
_____

215) 394
x 732
_____

216) 459
x 774
_____

217) 867
x 384
_____

218) 520
x 286
_____

219) 558
x 230
_____

220) 498
x 575
_____

221) 549
x 838
_____

222) 972
x 463
_____

223) 846
x 818
_____

224) 823
x 364
_____

225) 601
x 848
_____

# Multiplication

226) 582
x 963
_____

227) 652
x 275
_____

228) 377
x 876
_____

229) 641
x 626
_____

230) 190
x 579
_____

231) 697
x 224
_____

232) 315
x 757
_____

233) 423
x 273
_____

234) 984
x 637
_____

235) 982
x 385
_____

236) 248
x 348
_____

237) 451
x 785
_____

238) 935
x 847
_____

239) 755
x 812
_____

240) 115
x 485
_____

# Multiplication

241) 928
x 641

242) 361
x 788

243) 647
x 663

244) 996
x 235

245) 186
x 292

246) 378
x 637

247) 252
x 184

248) 573
x 947

249) 388
x 592

250) 507
x 617

251) 577
x 734

252) 522
x 736

253) 363
x 354

254) 925
x 347

255) 304
x 693

# Multiplication

256) 412
   x 295
_____

257) 901
   x 425
_____

258) 727
   x 846
_____

259) 340
   x 130
_____

260) 819
   x 318
_____

261) 926
   x 286
_____

262) 327
   x 121
_____

263) 535
   x 509
_____

264) 417
   x 214
_____

265) 584
   x 691
_____

266) 809
   x 225
_____

267) 790
   x 856
_____

268) 165
   x 713
_____

269) 733
   x 972
_____

270) 531
   x 548
_____

# Multiplication

271) 630
x 448
_____

272) 846
x 670
_____

273) 618
x 193
_____

274) 508
x 853
_____

275) 150
x 103
_____

276) 383
x 732
_____

277) 308
x 723
_____

278) 211
x 330
_____

279) 570
x 981
_____

280) 888
x 676
_____

281) 409
x 214
_____

282) 845
x 519
_____

283) 277
x 713
_____

284) 240
x 424
_____

285) 363
x 738
_____

# Multiplication

286) 910
x 352
_____

287) 297
x 320
_____

288) 195
x 659
_____

289) 517
x 785
_____

290) 979
x 786
_____

291) 524
x 773
_____

292) 713
x 446
_____

293) 517
x 588
_____

294) 852
x 159
_____

295) 720
x 674
_____

296) 586
x 916
_____

297) 504
x 721
_____

298) 461
x 322
_____

299) 329
x 945
_____

300) 204
x 382
_____

# Division

1)  870
    ÷ 391
    _____

2)  929
    ÷ 729
    _____

3)  931
    ÷ 828
    _____

4)  813
    ÷ 449
    _____

5)  300
    ÷ 271
    _____

6)  714
    ÷ 414
    _____

7)  778
    ÷ 381
    _____

8)  802
    ÷ 548
    _____

9)  862
    ÷ 582
    _____

10) 307
    ÷ 145
    _____

11) 700
    ÷ 180
    _____

12) 811
    ÷ 516
    _____

13) 816
    ÷ 675
    _____

14) 840
    ÷ 551
    _____

15) 689
    ÷ 143
    _____

16) 700
÷ 673
_____

17) 963
÷ 129
_____

18) 871
÷ 307
_____

19) 659
÷ 107
_____

20) 964
÷ 230
_____

21) 558
÷ 510
_____

22) 981
÷ 887
_____

23) 568
÷ 535
_____

24) 765
÷ 469
_____

25) 892
÷ 397
_____

26) 752
÷ 348
_____

27) 509
÷ 413
_____

28) 608
÷ 179
_____

29) 485
÷ 412
_____

30) 923
÷ 307
_____

# Division

31)  837
   ÷ 628
   _____

32)  468
   ÷ 242
   _____

33)  991
   ÷ 666
   _____

34)  999
   ÷ 704
   _____

35)  782
   ÷ 403
   _____

36)  707
   ÷ 118
   _____

37)  851
   ÷ 127
   _____

38)  744
   ÷ 137
   _____

39)  380
   ÷ 292
   _____

40)  598
   ÷ 250
   _____

41)  796
   ÷ 114
   _____

42)  733
   ÷ 269
   _____

43)  762
   ÷ 118
   _____

44)  762
   ÷ 282
   _____

45)  816
   ÷ 421
   _____

# Division

46) 272
÷ 164
_____

47) 779
÷ 121
_____

48) 943
÷ 693
_____

49) 662
÷ 136
_____

50) 435
÷ 206
_____

51) 486
÷ 141
_____

52) 821
÷ 168
_____

53) 760
÷ 321
_____

54) 398
÷ 365
_____

55) 521
÷ 471
_____

56) 621
÷ 374
_____

57) 341
÷ 213
_____

58) 566
÷ 146
_____

59) 313
÷ 281
_____

60) 738
÷ 696
_____

# Division

61) 639
   ÷ 445
   _____

62) 919
   ÷ 577
   _____

63) 985
   ÷ 211
   _____

64) 945
   ÷ 517
   _____

65) 767
   ÷ 583
   _____

66) 473
   ÷ 171
   _____

67) 909
   ÷ 293
   _____

68) 690
   ÷ 212
   _____

69) 999
   ÷ 294
   _____

70) 660
   ÷ 111
   _____

71) 872
   ÷ 606
   _____

72) 828
   ÷ 816
   _____

73) 360
   ÷ 222
   _____

74) 793
   ÷ 428
   _____

75) 572
   ÷ 405
   _____

# Division

76) 802
÷ 143
_____

77) 713
÷ 329
_____

78) 968
÷ 367
_____

79) 637
÷ 440
_____

80) 593
÷ 393
_____

81) 936
÷ 700
_____

82) 879
÷ 750
_____

83) 417
÷ 305
_____

84) 673
÷ 198
_____

85) 507
÷ 456
_____

86) 694
÷ 154
_____

87) 786
÷ 677
_____

88) 975
÷ 597
_____

89) 498
÷ 334
_____

90) 915
÷ 655
_____

# Division

91) 655
÷ 621
_____

92) 739
÷ 212
_____

93) 208
÷ 106
_____

94) 600
÷ 342
_____

95) 786
÷ 456
_____

96) 750
÷ 700
_____

97) 869
÷ 704
_____

98) 970
÷ 267
_____

99) 863
÷ 652
_____

100) 690
÷ 426
_____

101) 963
÷ 120
_____

102) 691
÷ 318
_____

103) 198
÷ 141
_____

104) 569
÷ 265
_____

105) 369
÷ 203
_____

**106)** 988
÷ 594
_____

**107)** 526
÷ 235
_____

**108)** 737
÷ 632
_____

**109)** 424
÷ 352
_____

**110)** 950
÷ 582
_____

**111)** 462
÷ 185
_____

**112)** 990
÷ 571
_____

**113)** 862
÷ 518
_____

**114)** 830
÷ 732
_____

**115)** 703
÷ 107
_____

**116)** 734
÷ 720
_____

**117)** 766
÷ 681
_____

**118)** 852
÷ 741
_____

**119)** 972
÷ 662
_____

**120)** 887
÷ 458
_____

# Division

121) 980
÷ 715
_____

122) 283
÷ 184
_____

123) 830
÷ 190
_____

124) 911
÷ 708
_____

125) 361
÷ 353
_____

126) 886
÷ 832
_____

127) 424
÷ 290
_____

128) 781
÷ 740
_____

129) 940
÷ 286
_____

130) 960
÷ 132
_____

131) 689
÷ 548
_____

132) 522
÷ 301
_____

133) 795
÷ 426
_____

134) 423
÷ 217
_____

135) 729
÷ 276
_____

# Division

136) 976 ÷ 247

137) 960 ÷ 191

138) 616 ÷ 374

139) 571 ÷ 184

140) 147 ÷ 102

141) 684 ÷ 292

142) 479 ÷ 342

143) 964 ÷ 788

144) 262 ÷ 259

145) 955 ÷ 523

146) 666 ÷ 420

147) 860 ÷ 564

148) 848 ÷ 302

149) 755 ÷ 551

150) 181 ÷ 165

# Division

**151)** 949
÷ 477
_____

**152)** 549
÷ 209
_____

**153)** 862
÷ 312
_____

**154)** 376
÷ 295
_____

**155)** 696
÷ 499
_____

**156)** 564
÷ 156
_____

**157)** 641
÷ 195
_____

**158)** 617
÷ 122
_____

**159)** 697
÷ 451
_____

**160)** 698
÷ 688
_____

**161)** 271
÷ 161
_____

**162)** 703
÷ 684
_____

**163)** 861
÷ 433
_____

**164)** 514
÷ 478
_____

**165)** 485
÷ 426
_____

# Division

166) 621
÷ 344
_____

167) 684
÷ 234
_____

168) 718
÷ 451
_____

169) 468
÷ 241
_____

170) 887
÷ 613
_____

171) 878
÷ 443
_____

172) 863
÷ 673
_____

173) 905
÷ 148
_____

174) 575
÷ 142
_____

175) 839
÷ 705
_____

176) 647
÷ 415
_____

177) 344
÷ 254
_____

178) 852
÷ 524
_____

179) 791
÷ 275
_____

180) 981
÷ 691
_____

# Division

181) 210
   ÷ 186
_____

182) 652
   ÷ 587
_____

183) 528
   ÷ 237
_____

184) 910
   ÷ 542
_____

185) 463
   ÷ 266
_____

186) 749
   ÷ 486
_____

187) 547
   ÷ 451
_____

188) 646
   ÷ 438
_____

189) 796
   ÷ 256
_____

190) 626
   ÷ 258
_____

191) 880
   ÷ 868
_____

192) 692
   ÷ 625
_____

193) 921
   ÷ 818
_____

194) 334
   ÷ 118
_____

195) 521
   ÷ 304
_____

# Division

**196)** 167
÷ 104
_____

**197)** 607
÷ 203
_____

**198)** 882
÷ 635
_____

**199)** 472
÷ 213
_____

**200)** 990
÷ 854
_____

**201)** 251
÷ 198
_____

**202)** 809
÷ 579
_____

**203)** 658
÷ 652
_____

**204)** 614
÷ 395
_____

**205)** 744
÷ 176
_____

**206)** 750
÷ 183
_____

**207)** 951
÷ 667
_____

**208)** 713
÷ 585
_____

**209)** 873
÷ 350
_____

**210)** 582
÷ 332
_____

# Division

211) 359
÷ 192
_____

212) 461
÷ 134
_____

213) 588
÷ 294
_____

214) 751
÷ 391
_____

215) 887
÷ 770
_____

216) 397
÷ 232
_____

217) 492
÷ 372
_____

218) 850
÷ 352
_____

219) 671
÷ 257
_____

220) 553
÷ 222
_____

221) 685
÷ 626
_____

222) 891
÷ 321
_____

223) 630
÷ 270
_____

224) 678
÷ 438
_____

225) 751
÷ 244
_____

# Division

226) 693
÷ 539
_____

227) 948
÷ 428
_____

228) 973
÷ 806
_____

229) 341
÷ 211
_____

230) 902
÷ 797
_____

231) 969
÷ 369
_____

232) 721
÷ 105
_____

233) 871
÷ 523
_____

234) 570
÷ 483
_____

235) 929
÷ 533
_____

236) 321
÷ 135
_____

237) 853
÷ 330
_____

238) 991
÷ 659
_____

239) 609
÷ 572
_____

240) 363
÷ 303
_____

# Division

241) 576
÷ 427
_____

242) 594
÷ 204
_____

243) 975
÷ 946
_____

244) 660
÷ 567
_____

245) 893
÷ 349
_____

246) 414
÷ 319
_____

247) 672
÷ 447
_____

248) 778
÷ 707
_____

249) 701
÷ 317
_____

250) 711
÷ 559
_____

251) 239
÷ 167
_____

252) 626
÷ 519
_____

253) 276
÷ 211
_____

254) 931
÷ 556
_____

255) 369
÷ 275
_____

# Division

256) 827
÷ 660
_____

257) 481
÷ 367
_____

258) 302
÷ 181
_____

259) 901
÷ 711
_____

260) 743
÷ 192
_____

261) 593
÷ 357
_____

262) 263
÷ 184
_____

263) 693
÷ 653
_____

264) 325
÷ 240
_____

265) 871
÷ 627
_____

266) 539
÷ 308
_____

267) 804
÷ 213
_____

268) 899
÷ 538
_____

269) 864
÷ 623
_____

270) 924
÷ 745
_____

# Division

271) 303
÷ 142
_____

272) 646
÷ 639
_____

273) 979
÷ 546
_____

274) 884
÷ 507
_____

275) 859
÷ 248
_____

276) 990
÷ 472
_____

277) 986
÷ 502
_____

278) 785
÷ 465
_____

279) 996
÷ 191
_____

280) 867
÷ 864
_____

281) 948
÷ 415
_____

282) 814
÷ 170
_____

283) 474
÷ 330
_____

284) 603
÷ 414
_____

285) 730
÷ 363
_____

**This is the last page, but there are 7 books in this series (Plus other series too!)...**

286) 886 ÷ 872

287) 995 ÷ 377

288) 705 ÷ 159

289) 871 ÷ 227

290) 712 ÷ 217

291) 738 ÷ 236

292) 584 ÷ 510

293) 634 ÷ 391

294) 422 ÷ 191

295) 709 ÷ 403

296) 544 ÷ 309

297) 470 ÷ 156

298) 360 ÷ 287

299) 562 ÷ 508

300) 622 ÷ 177

...Flip the page for bonus samples of the other books/series.

# Bonus: Next Book Sample – 4-Digit Problems

Enjoy this sample of *Math Book 3* which has 4-digit problems:

1)   2816
   + 9138
   _____

2)   5170
   + 1475
   _____

3)   7016
   + 5417
   _____

4)   1888
   - 1874
   _____

5)   9157
   - 8548
   _____

6)   3693
   - 1040
   _____

7)   6449
   x 5675
   _____

8)   8873
   x 3994
   _____

9)   8291
   x 9402
   _____

10)  2375
   ÷ 1967
   _____

11)  9122
   ÷ 6619
   _____

12)  3400
   ÷ 2575
   _____

**Math Book 3 (4-Digit Problems)**
Over 1000 problems in large print with solutions.

**Available on Amazon** Visit: smile.ws/pma9

# Bonus: Mixed Operations Sample

Enjoy this sample of *Math Book 7* which has mixed operation problems:

1) (11 - 8) X (9 - 19)

8) (2 X 8) + (6 X 11)

2) (16 - 14) + (4 X 2)

9) (9 + 5) - (0 - 3)

3) (15 X 4) + (3 X 8)

10) (5 + 9) + (5 - 6)

4) (17 + 18) - (3 X 4)

11) (14 X 7) - (12 X 1)

5) (8 + 12) X (15 - 7)

12) (18 - 10) X (14 X 9)

6) (7 - 13) - (14 + 6)

13) (7 X 3) + (19 X 10)

7) (2 + 5) - (1 + 12)

14) (1 - 6) + (4 - 10)

Answers for sample puzzles are in their respective full books.

# Bonus: Multi-Step Long Math Sample

A fun twist on math! Each problem has multiple steps.
Solve the first pair in each problem. Then take your answer and
use it with the next step in the same problem (and so on).

**55.** 799
353
- 187

**56.** 354
666
+ 547

**57.** 776
746
x  47

**94.** 693
273
83
x    9

**95.** 720
465
182
-   34

**96.** 940
791
783
+ 372

**115.** 905
444
181
-   10

**116.** 866
756
213
+ 779

**117.** 497
82
42
x    9

(Sample is non-sequential to reflect the variety in the book.)

# Bonus: Algebra Book Sample

## Solve for $x$

1) $-2x + 10 = 4$

2) $-15x + 16 = 16$

3) $-14x - 10 = 4$

4) $-4x + 12 = -13$

5) $x + 9 = -7$

6) $-13x + 11 = -13$

7) $13x + 4 = -12$

8) $-5x + 5 = -3$

9) $-3x - 1 = -2$

10) $3x + 12 = -14$

Answers for sample puzzles are in their respective full books.

# Addition: **Answers**

1) 624

2) 1,219

3) 966

4) 1,290

5) 1,546

6) 839

7) 1,150

8) 1,042

9) 548

10) 1,146

11) 1,549

12) 1,297

13) 947

14) 1,378

15) 1,375

16) 1,220

17) 786

18) 1,231

19) 858

20) 405

21) 1,776

22) 1,170

23) 1,113

24) 863

25) 981

26) 1,013

27) 948

28) 916

29) 787

30) 418

31) 1,473

32) 641

33) 1,179

34) 1,281

35) 1,148

36) 527

37) 1,680

38) 1,213

39) 1,160

40) 850

41) 1,301

42) 578

43) 793

44) 974

45) 1,346

46) 729

47) 1,181

48) 1,197

49) 722

50) 1,396

51) 845

52) 865

53) 982

54) 1,363

55) 811

56) 1,480

57) 1,097

58) 1,093

59) 1,050

60) 1,564

61) 838

62) 1,035

63) 1,474

64) 1,287

65) 1,852

66) 972

67) 1,660

68) 748

69) 928

70) 949

71) 986

72) 705

73) 1,202

74) 760

75) 793

76) 1,702

77) 1,579

78) 606

79) 646

80) 935

81) 1,485

82) 786

83) 735

84) 1,347

85) 1,400

86) 1,068

87) 308

88) 1,297

89) 601

90) 734

91) 1,058

92) 1,278

93) 950

94) 906

95) 1,163

96) 873

97) 1,911

98) 1,330

99) 1,488

100) 1,056

101) 1,091

102) 1,571

103) 686

104) 961

105) 1,196

106) 1,286

107) 846

108) 1,108

109) 1,468

110) 661

111) 1,526

112) 1,127

113) 358

114) 1,459

115) 1,197

116) 1,134

117) 951

118) 1,297

119) 1,589

120) 1,279

# Addition: Answers

| | | |
|---|---|---|
| **121)** 1,332 | **122)** 816 | **123)** 920 |
| **124)** 1,391 | **125)** 935 | **126)** 1,610 |
| **127)** 947 | **128)** 1,100 | **129)** 817 |
| **130)** 460 | **131)** 1,724 | **132)** 744 |
| **133)** 1,733 | **134)** 1,365 | **135)** 1,168 |
| **136)** 1,785 | **137)** 1,497 | **138)** 1,157 |
| **139)** 1,264 | **140)** 1,826 | **141)** 1,204 |
| **142)** 260 | **143)** 1,742 | **144)** 762 |
| **145)** 1,038 | **146)** 1,001 | **147)** 592 |
| **148)** 527 | **149)** 1,037 | **150)** 391 |
| **151)** 327 | **152)** 1,515 | **153)** 1,413 |
| **154)** 874 | **155)** 990 | **156)** 1,228 |
| **157)** 1,020 | **158)** 953 | **159)** 1,636 |
| **160)** 1,104 | **161)** 351 | **162)** 1,488 |
| **163)** 1,126 | **164)** 1,189 | **165)** 1,623 |
| **166)** 1,739 | **167)** 1,142 | **168)** 1,672 |
| **169)** 1,247 | **170)** 999 | **171)** 1,360 |
| **172)** 556 | **173)** 754 | **174)** 957 |
| **175)** 945 | **176)** 939 | **177)** 1,298 |
| **178)** 916 | **179)** 1,476 | **180)** 1,577 |

# Addition: Answers

181) 902        182) 744        183) 1,291

184) 501        185) 1,408        186) 1,115

187) 1,400        188) 1,105        189) 328

190) 1,389        191) 1,668        192) 1,246

193) 1,076        194) 820        195) 1,015

196) 1,260        197) 1,121        198) 833

199) 955        200) 1,324        201) 861

202) 1,151        203) 973        204) 680

205) 517        206) 853        207) 399

208) 1,774        209) 1,479        210) 1,578

211) 1,495        212) 1,374        213) 1,726

214) 1,489        215) 709        216) 936

217) 1,386        218) 1,902        219) 941

220) 968        221) 1,621        222) 1,193

223) 1,011        224) 1,797        225) 1,018

226) 824        227) 840        228) 1,280

229) 305        230) 1,423        231) 855

232) 1,134        233) 1,274        234) 818

235) 999        236) 554        237) 1,084

238) 1,137        239) 1,686        240) 1,082

# Addition: Answers

241) 1,014  242) 1,425  243) 1,093

244) 1,064  245) 1,068  246) 908

247) 1,275  248) 1,035  249) 966

250) 1,312  251) 1,376  252) 1,349

253) 1,244  254) 503  255) 1,205

256) 971  257) 1,536  258) 913

259) 716  260) 1,204  261) 1,473

262) 999  263) 1,845  264) 873

265) 621  266) 915  267) 1,162

268) 1,577  269) 1,450  270) 708

271) 1,062  272) 738  273) 1,473

274) 479  275) 790  276) 1,174

277) 309  278) 482  279) 1,822

280) 1,286  281) 1,233  282) 322

283) 1,147  284) 1,098  285) 1,364

286) 1,139  287) 1,256  288) 1,416

289) 1,655  290) 1,501  291) 1,379

292) 1,515  293) 1,047  294) 780

295) 1,009  296) 661  297) 1,067

298) 785  299) 843  300) 409

# Subtraction: **Answers**

| | | |
|---|---|---|
| 1) 356 | 2) 530 | 3) 290 |
| 4) 12 | 5) 295 | 6) 189 |
| 7) 49 | 8) 442 | 9) 111 |
| 10) 583 | 11) 734 | 12) 727 |
| 13) 541 | 14) 90 | 15) 163 |
| 16) 41 | 17) 115 | 18) 398 |
| 19) 413 | 20) 302 | 21) 46 |
| 22) 775 | 23) 368 | 24) 530 |
| 25) 326 | 26) 77 | 27) 203 |
| 28) 35 | 29) 523 | 30) 535 |
| 31) 91 | 32) 226 | 33) 19 |
| 34) 295 | 35) 110 | 36) 402 |
| 37) 435 | 38) 477 | 39) 180 |
| 40) 185 | 41) 64 | 42) 92 |
| 43) 472 | 44) 156 | 45) 279 |
| 46) 106 | 47) 141 | 48) 121 |
| 49) 47 | 50) 111 | 51) 213 |
| 52) 618 | 53) 167 | 54) 48 |
| 55) 95 | 56) 235 | 57) 132 |
| 58) 41 | 59) 257 | 60) 300 |

# Subtraction: **Answers**

| | | |
|---|---|---|
| **61)** 54 | **62)** 675 | **63)** 300 |
| **64)** 622 | **65)** 288 | **66)** 378 |
| **67)** 165 | **68)** 224 | **69)** 422 |
| **70)** 662 | **71)** 253 | **72)** 18 |
| **73)** 73 | **74)** 396 | **75)** 604 |
| **76)** 42 | **77)** 477 | **78)** 479 |
| **79)** 573 | **80)** 415 | **81)** 315 |
| **82)** 402 | **83)** 279 | **84)** 113 |
| **85)** 838 | **86)** 9 | **87)** 660 |
| **88)** 255 | **89)** 1 | **90)** 376 |
| **91)** 31 | **92)** 686 | **93)** 478 |
| **94)** 495 | **95)** 280 | **96)** 235 |
| **97)** 327 | **98)** 340 | **99)** 357 |
| **100)** 104 | **101)** 397 | **102)** 542 |
| **103)** 571 | **104)** 125 | **105)** 3 |
| **106)** 401 | **107)** 67 | **108)** 175 |
| **109)** 255 | **110)** 424 | **111)** 346 |
| **112)** 8 | **113)** 400 | **114)** 142 |
| **115)** 207 | **116)** 712 | **117)** 66 |
| **118)** 131 | **119)** 256 | **120)** 579 |

| | | |
|---|---|---|
| **121)** 449 | **122)** 313 | **123)** 138 |
| **124)** 293 | **125)** 578 | **126)** 464 |
| **127)** 399 | **128)** 527 | **129)** 77 |
| **130)** 806 | **131)** 511 | **132)** 215 |
| **133)** 203 | **134)** 300 | **135)** 81 |
| **136)** 180 | **137)** 236 | **138)** 168 |
| **139)** 27 | **140)** 87 | **141)** 21 |
| **142)** 383 | **143)** 1 | **144)** 406 |
| **145)** 425 | **146)** 95 | **147)** 265 |
| **148)** 3 | **149)** 500 | **150)** 26 |
| **151)** 230 | **152)** 410 | **153)** 66 |
| **154)** 180 | **155)** 133 | **156)** 637 |
| **157)** 155 | **158)** 598 | **159)** 138 |
| **160)** 443 | **161)** 332 | **162)** 398 |
| **163)** 317 | **164)** 120 | **165)** 106 |
| **166)** 45 | **167)** 80 | **168)** 279 |
| **169)** 52 | **170)** 111 | **171)** 488 |
| **172)** 75 | **173)** 94 | **174)** 675 |
| **175)** 564 | **176)** 36 | **177)** 286 |
| **178)** 248 | **179)** 254 | **180)** 21 |

# Subtraction: **Answers**

| | | |
|---|---|---|
| **181)** 793 | **182)** 185 | **183)** 505 |
| **184)** 669 | **185)** 696 | **186)** 5 |
| **187)** 216 | **188)** 182 | **189)** 139 |
| **190)** 32 | **191)** 382 | **192)** 789 |
| **193)** 10 | **194)** 305 | **195)** 26 |
| **196)** 312 | **197)** 500 | **198)** 41 |
| **199)** 221 | **200)** 3 | **201)** 529 |
| **202)** 539 | **203)** 368 | **204)** 413 |
| **205)** 257 | **206)** 511 | **207)** 633 |
| **208)** 596 | **209)** 525 | **210)** 572 |
| **211)** 361 | **212)** 63 | **213)** 477 |
| **214)** 142 | **215)** 676 | **216)** 200 |
| **217)** 88 | **218)** 439 | **219)** 618 |
| **220)** 259 | **221)** 702 | **222)** 118 |
| **223)** 192 | **224)** 12 | **225)** 604 |
| **226)** 27 | **227)** 641 | **228)** 94 |
| **229)** 69 | **230)** 240 | **231)** 54 |
| **232)** 726 | **233)** 382 | **234)** 119 |
| **235)** 275 | **236)** 55 | **237)** 207 |
| **238)** 364 | **239)** 459 | **240)** 382 |

| | | |
|---|---|---|
| 241) 334 | 242) 60 | 243) 130 |
| 244) 521 | 245) 411 | 246) 761 |
| 247) 166 | 248) 475 | 249) 150 |
| 250) 177 | 251) 54 | 252) 257 |
| 253) 392 | 254) 573 | 255) 230 |
| 256) 436 | 257) 387 | 258) 81 |
| 259) 831 | 260) 31 | 261) 481 |
| 262) 685 | 263) 617 | 264) 80 |
| 265) 349 | 266) 69 | 267) 80 |
| 268) 150 | 269) 680 | 270) 485 |
| 271) 531 | 272) 399 | 273) 0 |
| 274) 325 | 275) 463 | 276) 237 |
| 277) 232 | 278) 73 | 279) 55 |
| 280) 300 | 281) 236 | 282) 3 |
| 283) 253 | 284) 251 | 285) 582 |
| 286) 370 | 287) 386 | 288) 31 |
| 289) 245 | 290) 222 | 291) 681 |
| 292) 82 | 293) 426 | 294) 131 |
| 295) 295 | 296) 26 | 297) 78 |
| 298) 207 | 299) 68 | 300) 534 |

# Multiplication: Answers

1) 70,560

2) 131,567

3) 736,139

4) 226,526

5) 68,628

6) 516,957

7) 85,044

8) 532,800

9) 142,065

10) 130,494

11) 242,352

12) 521,040

13) 99,604

14) 648,639

15) 323,750

16) 236,621

17) 77,945

18) 341,600

19) 252,300

20) 156,350

21) 303,620

22) 66,096

23) 34,648

24) 176,784

25) 191,565

26) 568,066

27) 829,821

28) 203,801

29) 317,504

30) 66,757

31) 59,058

32) 147,520

33) 132,498

34) 373,932

35) 431,218

36) 396,720

37) 498,225

38) 309,720

39) 167,520

40) 355,885

41) 66,415

42) 870,381

43) 29,766

44) 413,660

45) 131,044

46) 275,158

47) 303,316

48) 48,600

49) 769,984

50) 100,344

51) 224,439

52) 207,900

53) 484,866

54) 218,500

55) 214,786

56) 134,850

57) 190,026

58) 84,410

59) 237,882

60) 28,363

# Multiplication: **Answers**

61) 158,238

62) 688,257

63) 126,474

64) 65,772

65) 832,795

66) 442,330

67) 149,224

68) 484,584

69) 82,950

70) 61,105

71) 142,913

72) 388,068

73) 337,260

74) 240,036

75) 363,312

76) 302,670

77) 131,793

78) 135,786

79) 41,520

80) 684,440

81) 145,452

82) 263,412

83) 639,665

84) 265,088

85) 342,916

86) 202,840

87) 320,705

88) 237,088

89) 47,565

90) 458,948

91) 479,385

92) 367,600

93) 182,672

94) 183,996

95) 650,934

96) 162,146

97) 368,180

98) 75,808

99) 688,014

100) 440,016

101) 48,870

102) 80,408

103) 386,325

104) 732,683

105) 540,690

106) 251,563

107) 205,570

108) 119,358

109) 432,825

110) 398,792

111) 548,576

112) 377,853

113) 128,443

114) 130,680

115) 268,092

116) 147,225

117) 68,400

118) 119,364

119) 583,725

120) 300,530

# Multiplication: Answers

121) 241,087     122) 279,072     123) 376,248

124) 471,653     125) 734,160     126) 85,629

127) 254,043     128) 90,706      129) 440,847

130) 652,026     131) 466,395     132) 165,509

133) 339,427     134) 68,474      135) 84,609

136) 98,975      137) 353,612     138) 354,750

139) 395,038     140) 364,770     141) 157,542

142) 148,953     143) 60,928      144) 57,856

145) 101,814     146) 499,016     147) 107,646

148) 542,982     149) 61,880      150) 62,167

151) 435,664     152) 866,768     153) 131,895

154) 583,296     155) 39,269      156) 807,363

157) 75,520      158) 106,173     159) 108,400

160) 296,009     161) 515,372     162) 63,080

163) 111,048     164) 63,042      165) 319,032

166) 82,669      167) 707,826     168) 83,452

169) 523,342     170) 612,444     171) 304,194

172) 909,090     173) 112,398     174) 279,450

175) 76,293      176) 383,400     177) 163,863

178) 328,520     179) 164,897     180) 460,356

# Multiplication: **Answers**

181) 142,819    182) 363,268    183) 426,474

184) 645,300    185) 557,244    186) 195,394

187) 584,288    188) 92,800     189) 97,848

190) 439,112    191) 171,740    192) 561,724

193) 145,134    194) 37,380     195) 68,975

196) 128,679    197) 824,934    198) 467,441

199) 575,043    200) 385,475    201) 93,555

202) 343,824    203) 212,850    204) 755,016

205) 370,240    206) 321,305    207) 576,720

208) 61,506     209) 275,912    210) 196,571

211) 142,393    212) 646,256    213) 126,160

214) 122,836    215) 288,408    216) 355,266

217) 332,928    218) 148,720    219) 128,340

220) 286,350    221) 460,062    222) 450,036

223) 692,028    224) 299,572    225) 509,648

226) 560,466    227) 179,300    228) 330,252

229) 401,266    230) 110,010    231) 156,128

232) 238,455    233) 115,479    234) 626,808

235) 378,070    236) 86,304     237) 354,035

238) 791,945    239) 613,060    240) 55,775

# Multiplication: **Answers**

| | | |
|---|---|---|
| **241)** 594,848 | **242)** 284,468 | **243)** 428,961 |
| **244)** 234,060 | **245)** 54,312 | **246)** 240,786 |
| **247)** 46,368 | **248)** 542,631 | **249)** 229,696 |
| **250)** 312,819 | **251)** 423,518 | **252)** 384,192 |
| **253)** 128,502 | **254)** 320,975 | **255)** 210,672 |
| **256)** 121,540 | **257)** 382,925 | **258)** 615,042 |
| **259)** 44,200 | **260)** 260,442 | **261)** 264,836 |
| **262)** 39,567 | **263)** 272,315 | **264)** 89,238 |
| **265)** 403,544 | **266)** 182,025 | **267)** 676,240 |
| **268)** 117,645 | **269)** 712,476 | **270)** 290,988 |
| **271)** 282,240 | **272)** 566,820 | **273)** 119,274 |
| **274)** 433,324 | **275)** 15,450 | **276)** 280,356 |
| **277)** 222,684 | **278)** 69,630 | **279)** 559,170 |
| **280)** 600,288 | **281)** 87,526 | **282)** 438,555 |
| **283)** 197,501 | **284)** 101,760 | **285)** 267,894 |
| **286)** 320,320 | **287)** 95,040 | **288)** 128,505 |
| **289)** 405,845 | **290)** 769,494 | **291)** 405,052 |
| **292)** 317,998 | **293)** 303,996 | **294)** 135,468 |
| **295)** 485,280 | **296)** 536,776 | **297)** 363,384 |
| **298)** 148,442 | **299)** 310,905 | **300)** 77,928 |

# Division: **Answers**

| | | |
|---|---|---|
| **1)** 2.23 | **2)** 1.27 | **3)** 1.12 |
| **4)** 1.81 | **5)** 1.11 | **6)** 1.72 |
| **7)** 2.04 | **8)** 1.46 | **9)** 1.48 |
| **10)** 2.12 | **11)** 3.89 | **12)** 1.57 |
| **13)** 1.21 | **14)** 1.52 | **15)** 4.82 |
| **16)** 1.04 | **17)** 7.47 | **18)** 2.84 |
| **19)** 6.16 | **20)** 4.19 | **21)** 1.09 |
| **22)** 1.11 | **23)** 1.06 | **24)** 1.63 |
| **25)** 2.25 | **26)** 2.16 | **27)** 1.23 |
| **28)** 3.40 | **29)** 1.18 | **30)** 3.01 |
| **31)** 1.33 | **32)** 1.93 | **33)** 1.49 |
| **34)** 1.42 | **35)** 1.94 | **36)** 5.99 |
| **37)** 6.70 | **38)** 5.43 | **39)** 1.30 |
| **40)** 2.39 | **41)** 6.98 | **42)** 2.72 |
| **43)** 6.46 | **44)** 2.70 | **45)** 1.94 |
| **46)** 1.66 | **47)** 6.44 | **48)** 1.36 |
| **49)** 4.87 | **50)** 2.11 | **51)** 3.45 |
| **52)** 4.89 | **53)** 2.37 | **54)** 1.09 |
| **55)** 1.11 | **56)** 1.66 | **57)** 1.60 |
| **58)** 3.88 | **59)** 1.11 | **60)** 1.06 |

# Division: **Answers**

| | | |
|---|---|---|
| **61)** 1.44 | **62)** 1.59 | **63)** 4.67 |
| **64)** 1.83 | **65)** 1.32 | **66)** 2.77 |
| **67)** 3.10 | **68)** 3.25 | **69)** 3.40 |
| **70)** 5.95 | **71)** 1.44 | **72)** 1.01 |
| **73)** 1.62 | **74)** 1.85 | **75)** 1.41 |
| **76)** 5.61 | **77)** 2.17 | **78)** 2.64 |
| **79)** 1.45 | **80)** 1.51 | **81)** 1.34 |
| **82)** 1.17 | **83)** 1.37 | **84)** 3.40 |
| **85)** 1.11 | **86)** 4.51 | **87)** 1.16 |
| **88)** 1.63 | **89)** 1.49 | **90)** 1.40 |
| **91)** 1.05 | **92)** 3.49 | **93)** 1.96 |
| **94)** 1.75 | **95)** 1.72 | **96)** 1.07 |
| **97)** 1.23 | **98)** 3.63 | **99)** 1.32 |
| **100)** 1.62 | **101)** 8.03 | **102)** 2.17 |
| **103)** 1.40 | **104)** 2.15 | **105)** 1.82 |
| **106)** 1.66 | **107)** 2.24 | **108)** 1.17 |
| **109)** 1.20 | **110)** 1.63 | **111)** 2.50 |
| **112)** 1.73 | **113)** 1.66 | **114)** 1.13 |
| **115)** 6.57 | **116)** 1.02 | **117)** 1.12 |
| **118)** 1.15 | **119)** 1.47 | **120)** 1.94 |

| | | |
|---|---|---|
| **121)** 1.37 | **122)** 1.54 | **123)** 4.37 |
| **124)** 1.29 | **125)** 1.02 | **126)** 1.06 |
| **127)** 1.46 | **128)** 1.06 | **129)** 3.29 |
| **130)** 7.27 | **131)** 1.26 | **132)** 1.73 |
| **133)** 1.87 | **134)** 1.95 | **135)** 2.64 |
| **136)** 3.95 | **137)** 5.03 | **138)** 1.65 |
| **139)** 3.10 | **140)** 1.44 | **141)** 2.34 |
| **142)** 1.40 | **143)** 1.22 | **144)** 1.01 |
| **145)** 1.83 | **146)** 1.59 | **147)** 1.52 |
| **148)** 2.81 | **149)** 1.37 | **150)** 1.10 |
| **151)** 1.99 | **152)** 2.63 | **153)** 2.76 |
| **154)** 1.27 | **155)** 1.39 | **156)** 3.62 |
| **157)** 3.29 | **158)** 5.06 | **159)** 1.55 |
| **160)** 1.01 | **161)** 1.68 | **162)** 1.03 |
| **163)** 1.99 | **164)** 1.08 | **165)** 1.14 |
| **166)** 1.81 | **167)** 2.92 | **168)** 1.59 |
| **169)** 1.94 | **170)** 1.45 | **171)** 1.98 |
| **172)** 1.28 | **173)** 6.11 | **174)** 4.05 |
| **175)** 1.19 | **176)** 1.56 | **177)** 1.35 |
| **178)** 1.63 | **179)** 2.88 | **180)** 1.42 |

# Division: **Answers**

| | | |
|---|---|---|
| **181)** 1.13 | **182)** 1.11 | **183)** 2.23 |
| **184)** 1.68 | **185)** 1.74 | **186)** 1.54 |
| **187)** 1.21 | **188)** 1.47 | **189)** 3.11 |
| **190)** 2.43 | **191)** 1.01 | **192)** 1.11 |
| **193)** 1.13 | **194)** 2.83 | **195)** 1.71 |
| **196)** 1.61 | **197)** 2.99 | **198)** 1.39 |
| **199)** 2.22 | **200)** 1.16 | **201)** 1.27 |
| **202)** 1.40 | **203)** 1.01 | **204)** 1.55 |
| **205)** 4.23 | **206)** 4.10 | **207)** 1.43 |
| **208)** 1.22 | **209)** 2.49 | **210)** 1.75 |
| **211)** 1.87 | **212)** 3.44 | **213)** 2.00 |
| **214)** 1.92 | **215)** 1.15 | **216)** 1.71 |
| **217)** 1.32 | **218)** 2.41 | **219)** 2.61 |
| **220)** 2.49 | **221)** 1.09 | **222)** 2.78 |
| **223)** 2.33 | **224)** 1.55 | **225)** 3.08 |
| **226)** 1.29 | **227)** 2.21 | **228)** 1.21 |
| **229)** 1.62 | **230)** 1.13 | **231)** 2.63 |
| **232)** 6.87 | **233)** 1.67 | **234)** 1.18 |
| **235)** 1.74 | **236)** 2.38 | **237)** 2.58 |
| **238)** 1.50 | **239)** 1.06 | **240)** 1.20 |

241) 1.35

242) 2.91

243) 1.03

244) 1.16

245) 2.56

246) 1.30

247) 1.50

248) 1.10

249) 2.21

250) 1.27

251) 1.43

252) 1.21

253) 1.31

254) 1.67

255) 1.34

256) 1.25

257) 1.31

258) 1.67

259) 1.27

260) 3.87

261) 1.66

262) 1.43

263) 1.06

264) 1.35

265) 1.39

266) 1.75

267) 3.77

268) 1.67

269) 1.39

270) 1.24

271) 2.13

272) 1.01

273) 1.79

274) 1.74

275) 3.46

276) 2.10

277) 1.96

278) 1.69

279) 5.21

280) 1.00

281) 2.28

282) 4.79

283) 1.44

284) 1.46

285) 2.01

286) 1.02

287) 2.64

288) 4.43

289) 3.84

290) 3.28

291) 3.13

292) 1.15

293) 1.62

294) 2.21

295) 1.76

296) 1.76

297) 3.01

298) 1.25

299) 1.11

300) 3.51

# Math for all ages!

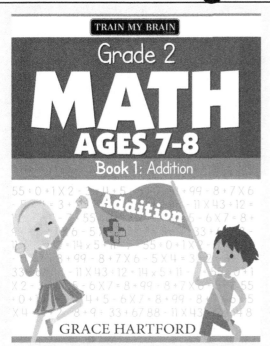

## Grade 2 Math
Math for ages 7 to 8
Visit: **smile.ws/pma8**

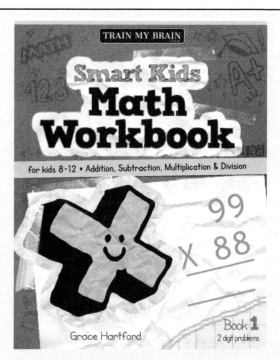

## Smart Kids Math
Math for ages 8 to 12
Visit: **smile.ws/pma5**

## Sudoku Series
Challenging number-based game
Visit: **smile.ws/ps9**

## Plus More Puzzle Books

Word Search, Word Scramble, Sudoku, Number Search, Mazes, and more – Enjoy a growing collection of puzzle books to train your brain!

**See them all at:**
**smile.ws/puzzles**

**Join our FREE club to get future books at discounted prices:**

**Visit: smile.ws/cma9**

Printed in Great Britain
by Amazon

32292917R00064